Edna Dantas Pereira

Productivity of melon fruit under osmopriming

Edna Dantas Pereira

Productivity of melon fruit under osmopriming

Melon fruit under seed conditioning

ScienciaScripts

Imprint

Cover image: www.ingimage.com

This book is a translation from the original published under ISBN 978-613-9-67556-2.

Publisher:
Sciencia Scripts
is a trademark of
Dodo Books Indian Ocean Ltd. and OmniScriptum S.R.L publishing group

120 High Road, East Finchley, London, N2 9ED, United Kingdom
Str. Armeneasca 28/1, office 1, Chisinau MD-2012, Republic of Moldova, Europe
Printed at: see last page
ISBN: 978-620-8-16914-5

SUMMARY

THANKS

To God, for everything he has done for me, especially for giving his son Jesus as a sacrifice in my place, and for allowing me to exist and reach the end of yet another phase of life;

To my family, especially my daughter Lorena Dantas Pinto for her immense love, support, affection and patience towards me;

In memory of my parents, Veneziano Pereira da Silva and Edenizia Macena Pereira, for their example of courage, perseverance and honesty, which gave all of us children the strength to carry on the fight and overcome the barriers inherent in life with credibility;

To my advisor D.Sc. Roberto Cleiton Fernandes de Queiroga, for the opportunity, patience and dedication;

To the examining board, Professors Marcos Eric Barbosa Brito and Joserlan Nonato Moreira, for their suggestions and corrections;

Professor Caciana Cavalcanti Costa, for her guidance during my teaching internship.

To the Federal University of Campina Grande, the Center for Agrifood Sciences and Technology and the Coordination of the Postgraduate Course in Tropical Horticulture, for the opportunities and the space provided for the development of this research;

To CNPq and CAPES for their contributions, without which it would not have been possible to carry out this work.

To all my teammates for contributing to this journey of work and research, especially Zaqueu Lopes, Rayana Ferreira Pereira, Lamartine Eduardo, Auderlan de Macena Pereira and Malbah Sena;

And finally, to all those who directly and indirectly contributed to this work and research.

SUMMARY

PEREIRA, Edna Dantas. Productivity and quality of melon fruits under Osmoconditioning and water salinity. 2016. 50f. Dissertation (Master's Degree in Tropical Horticulture) - Federal University of Campina Grande (UFCG), Pombal - PB, 2016.

Growing areas with intense evaporation, poor drainage and inadequate use of fertilizers have increased problems with salinity, causing damage to crop yields. The aim was to evaluate the productivity and quality of Cantaloupe melon fruit as a function of seed osmoconditioning and water salinity levels. The experiment was carried out in 2 x 4 subdivided plots in a randomized block design with four replications. The plots consisted of 4 levels of salinity in the irrigation water (0.3, 1.0, 1.7 and 2.4 dS m^{-1}) and the subplots of plants from seeds treated and untreated with saline NaCl solution. Characteristics related to melon fruit production and quality were assessed. The cultivar used was Rangers of the Cantaloupe group, planted at a spacing of 2.0 x 0.4m. The osmoconditioning of the seeds in NaCl solution affected the production components of the melon tree, where fruit mass was higher in plants from seeds treated with NaCl. In terms of productivity, it was clear that fruit mass contributed more than the number of fruits per plant and that this difference was only significant under low water salinity conditions of 0.3 dS m^{-1} . The quality of melon fruit was not altered by osmoconditioning the seed with NaCl and plants irrigated with water with a salinity of 2.4 dS m^{-1} reduced their production and quality of melon fruit.

Keywords: *Cucumis melo* L, pre-germination treatment, yield and soluble solids.

1 INTRODUCTION

The melon tree (*Cucumis melo* L.) is an olive belonging to the Cucurbitaceae family that produces fruit that is appreciated around the world and stands out on both the national and international markets due to its taste and high nutritional value. It is currently one of Brazil's most exported fresh fruits (NASCIMENTO NETO *et al.*, 2012).

Among fruit vegetables, the melon is the most prominent in the Northeast region of Brazil due to the favorable soil and climate conditions, especially with regard to the high temperature and high radiation that contributes to high evapotranspiration rates and low rainfall indices. These conditions are essential for the proper development and growth of plants and fruit, favoring increased production (ALVES, 2012).

In order to produce these fruits in this region, irrigation is necessary. However, the availability of low-salinity water is expensive with limited exploitable volume, while water from shallow wells is less expensive and has greater potential for use, although it can have a higher concentration of salts, which can reduce crop yields. In these areas, salinization is exacerbated by intense evaporation, poor drainage and the use of fertilizers, which affects crop yields (MEDEIROS *et al*, 2011).

The effects of salts on plants can be seen in the difficulty in absorbing saline water, the interference of salts in physiological processes, or even toxicity, similar to that caused by excessive fertilization (QUEIROGA *et al.*, 2006). With appropriate agricultural practices, plants will be able to assimilate tolerance when they suffer salt stress.

Studying the effects of salts on plants is essential for improving the production system by using appropriate production technologies for the region. Research is therefore needed to assess, for example, plant tolerance to the saline conditions found in production areas, with a view to increasing productivity and fruit quality at harvest (BESSA *et al.*, 2011).

One technology that can be used to economically produce under saline conditions is seed preparation through osmoconditioning. Physiological conditioning has been used to reduce germination time, increase germinability, uniformity and seedling vigor. Among the physiological conditioning techniques, osmoconditioning has been the most widely used. This technique consists of pre-treatment, in which the seeds are immersed in an osmotic solution at a specific time and temperature (PEREIRA *et al.*, 2008).

Sivitrepe *et al.* (2003) state that, in general, plants do not develop tolerance to salts unless they grow in saline conditions. According to these authors, high levels of salts in water reduce the growth rate of plants, resulting in smaller and fewer leaves.

With the aim of studying physiological conditioning procedures and verifying their relationship with the performance of seeds and plants in the cucumber field, Lima & Marcos Filho (2009) found that osmoconditioning on paper soaked in a solution of PEG -0.2 MPa proved to be efficient, more clearly affecting the speed than the percentage of germination of the batches evaluated. They also found that physiological conditioning enables the formation of vigorous cucumber seedlings, although these benefits do not extend to the production of fruit in the field.

This method has been used more in the production of vegetable seedlings and little is known about its effects on production. Thus, we can infer that the technique of physiological conditioning of seeds could bring progress in the production and quality of melon fruit through greater resistance of the seedling in the field when exposed to environmental adversities such as water or soil salinity.

The aim was to evaluate the yield and quality of Cantaloupe melon fruit as a function of seed osmoconditioning and water salinity levels.

2 LITERATURE REVIEW

2.1 Growing melons

The melon *(Cucumis melo* L.) originates from tropical African and Asian regions and, as such, conditions of high temperature and low relative humidity favor the best development of the plants, as well as providing high productivity and fruit of excellent quality (GRANGEIRO *et al.*, 2002).

The melon tree was commercially planted in Brazil in the mid-1960s, in the states of Rio Grande do Sul and Sao Paulo. During this period, the domestic market was supplied almost entirely by imports of fruit from Chile and Spain (ARAÙJO & VILELA, 2002).

The South and Southeast regions soon lost their prominence in melon production due to environmental factors that limited productivity and product quality (DIAS *et al.*, 1998). In recent decades, Brazil has gone from being an importer to an exporter of this vegetable, mainly due to production in the Northeast (FONTES & PUIATTI, 2005).

Melon cultivation in Brazil has grown considerably in recent years, mainly due to the good climatic conditions for crop development, such as high luminosity, low rainfall and low relative humidity. These environmental factors allow this crop to be produced all year round, favoring increased exports and ensuring economic and social expression, as direct jobs are generated in the various sectors of the production chain (MENDONÇA JÙNIOR, 2015).

Melon production in Brazil is concentrated almost entirely in the northeastern states, which together account for approximately 95% of national production. The main producing states are Rio Grande do Norte (254,530 t), Cearà (212,362 t), Bahia (33,431 t) and Pernambuco (20,410 t) (IBGE, 2014).

The economic importance of the crop has stimulated intensified research into the physiology, biochemistry (MOING *et al.*, 2010) and post-harvest preservation of the fruit (SENHOR *et al.*, 2008; TOMAZ *et al.*, 2009). Thus, melon is a tropical fruit of great commercial importance for Brazilian exports, and in 2011 it was the most exported fruit, with a volume of 169.6 thousand tons, representing more than half of total exports, with revenues of US$ 128.3 million (AGRIANUAL, 2012). This is largely due to the good acceptance of melons by the foreign market, especially by European countries.

The melon of the inodorus group is the most widely grown in the Northeast, accompanied by melons of the Cantaloupensis group, which are considered noble and have more attractive market characteristics (visual appearance, flavor, aroma, soluble solids and market value), representing 40% of the total cultivated area, leading to an increase in production, especially in the states of Rio Grande do Norte and Cearà (ALVES, 2012, MOREIRA et al., 2010). It is worth noting that these states are privileged because they are free of the cucurbit fly (*Anastreph grandis*), and marketing is favored due to the peak of their harvest occurring from September to January, which coincides with the world off-season (SENAR, 2007).

The melon fruit is eaten fresh, as an ingredient in salads with fruit or other vegetables, and in the form of juice, providing the human body with nutritional value such as vitamins and mineral salts, especially potassium, sodium and phosphorus. The energy value is relatively low, at 62 kcal $100g^{-1}$ of pulp. The total soluble solids content can reach up to 18% (ELMSTROM & MAYNARD, 1992). According to Bianco & Pratt (1977) apud Costa & Grangeiro (2003), over 97% of total soluble solids are sugars, with sucrose accounting for 50% of total soluble sugars.

The ripe fruit has medicinal properties and is considered calming, refreshing, alkalizing, mineralizing, oxidizing, diuretic, laxative and emollient. It is also recommended for controlling gout, rheumatism, arthritis, obesity, colitis, intestinal atony, constipation, kidney disorders, renal lithiasis, nephritis, cystitis, leucorrhea, urethritis and blenorrhagia (BALBACK, n.d., apud COSTA & GRANGEIRO, 2003).

Today, melon cultivars can be found in different regions, and this range is the result of great genetic variability, which has allowed different types of melon to adapt to different conditions (NUNES *et al.*, 2004), so that we can find fruit with varying characteristics depending on the group and cultivar.

However, in addition to large-scale producers, the cultivation of this vegetable has been expanding into small family farming areas as it shows significant and promising results in terms of economic return for producers, with the guarantee of consumer preference for melons. With this in mind, the aim is to learn about the levels of water salinity that each melon cultivar can tolerate, thus helping to improve management and minimize the effects of salts in the soil, as well as increasing crop productivity.

In recent years, noble melon cultivars, such as those from the Gàlia and Cantaloupe groups, which have more attractive organoleptic characteristics and a higher commercial value, have increased their market share by 15 to 20%. However, they are more sensitive, requiring more advanced cultivation techniques. Cantaloupe melons have also grown in production to serve the domestic market, due to their greater acceptance and the few restrictions imposed on their commercialization, especially for the regional market (MEDEIROS *et al.*, 2011).

2.2 Osmotic conditioning of seeds

The rapid and uniform establishment of seedlings in the field is fundamental to achieving the ideal stand and guaranteeing the desired productivity and quality of the final product. Osmotic conditioning, also known as physiological seed conditioning, priming or osmoconditioning (BITTENCOURT et al., 2004; SUNE, FRANKE; SAMPAIO, 2002), is a promising method for accelerating germination and allowing seedlings to emerge more quickly and uniformly in the field. This technique has become an ally in the production systems of various vegetable species, such as eggplant (REIS et al., 2013), cauliflower (MARCOS FILHO; KIKUTI; LIMA 2008), carrot (NASCIMENTO et al., 2009), onion (SINGH et al., 2014), cucumber (LIMA; MARCOS FILHO, 2010), tomato (AMOOAGHAIE; NIKZAD; SHAREGHI, 2010; ASGHARIPPOUR; RAFIEI, 2011) and okra (PANDITA et al., 2010; RAI; BASU, 2014), making it consistently essential for most seed companies.

The work of Heydecker, Higgins and Gulliver (1973) pioneered the development of the osmotic conditioning technique, which aims to allow faster root emergence when sowing, as well as synchronizing seedling emergence. In this way, the plants establish themselves more quickly and uniformly in the field, as well as minimizing the effects of unfavourable environmental conditions, due to less exposure of the seeds, attacks by pests and microorganisms, and providing greater competitiveness with weeds.

Various osmotic agents can be used for osmotic conditioning of seeds, such as inorganic salts like potassium nitrate (KNO_3), magnesium sulphate ($MgSO_4$), sodium chloride (NaCl), magnesium chloride ($MgCl_2$) and organic compounds like mannitol, glycerol and polyethylene glycol (PEG) (MARCOS FILHO, 2005).

Physiological conditioning, in general, can favor important biochemical events for seed germination, including active cell division during seedling growth (WATERWORTH *et al.*, 2011), DNA repair and replication, which are highly affected during the seed dehydration process, and compromise the synthesis of enzymes essential to the germination process (BEWLEY, 2009; CHEN; ARORA, 2013).

Osmoconditioning has provided promising results in terms of improving the percentage and speed of germination, emergence and vigor for various species, especially vegetables such as carrots (PEREIRA *et al.*, 2008), eggplants (GOMES *et al.*, 2012) and parsley (RODRIGUES *et al.*, 2009).

The percentage of germination in saline media is one of the most widespread methods for determining plant tolerance to excess salts. The decrease in germination potential and reduction in seedling vigor when subjected to saline concentrations compared to the control are indicative of the species' tolerance to salinity (OLIVEIRA *et al.*, 2008).

Using osmoconditioning with saline water conditions the seeds so that they can develop without further damage to their development and consequently less loss of germination, with better adaptation during their development. According to Tester & Davanporte, 2003, the response of plants to salinity can vary according to the species, cultivar and between phenological stages of the same genotype.

However, once the osmotic solution has been removed, the seeds can be transferred to places where they can germinate or be dried until they reach their original water content and then stored (NASCIMENTO; WEST, 2000; BIRUEL *et. al.*, 2007).

Pre-germination treatment of melon seeds with saline solution (6 g/L) resulted in a greater dry mass of the aerial part of the seedlings and a significant increase in the germination percentage; on the other hand, the salinity of the irrigation water reduced the leaf area, the dry mass of the aerial part of the seedling and the height of the seedling (QUEIROGA et al., 2006).

The development of these new technologies to elucidate the effects of physiological conditioning of seeds has contributed to safe decision-making in production systems, by providing subsidies that meet the demand for knowledge.

2.3 Water quality for irrigation

In Brazil and around the world, the issue of managing the planet's water is much debated. Irrigation is generally seen as the main tool to help produce the food that the growing and future population will demand (VENANCIO *et al.*, 2015).

Water is a limiting factor for agricultural development, and both lack and excess affect plant growth, health and production (PIRES *et al.,* 2001).

The lack of water resulting from the water crisis has had a number of direct effects on agricultural production, as well as reducing crop yields and product quality. As well as the quality of the water used for irrigation, neglecting this aspect has undesirable effects on crop management (MARÇAL, 2011). Water quality refers to its suitability for a particular use, i.e. whether its physical, chemical and biological characteristics meet the needs of the user (LIMA *et al.*, 1999).

Therefore, in various regions of the planet where rainfall is irregular, irrigation techniques play an important role in increasing agricultural productivity and, consequently, food production. The increase in the demand for water resources, with the expansion of irrigated areas, has been the subject of constant concern when aiming for sustainable development combined with the rational use of natural resources (REIS *et al.*, 2011).

According to Melo (2005), water quality is difficult to define because it is both simple and complete, given the complexity of the determining factors and the wide range of variable options available to describe irrigation water conditions. The existing quality is a function of the natural conditions and the occupation and use of the land in the hydrographic basin, while the desired quality is linked to the intended use of the water (VON SPERLING, 2005).

The study of the quality of water for irrigation should consider the quantification of electrical conductivity, as it is the measure of the capacity of the ions present in the water to conduct electricity and grows proportionally to the increase in the concentration of salts (RIBEIRO *et al.*, 2005). However, some important aspects need to be considered, such as salinity, sodicity, ion toxicity and bicarbonate levels. These variables are fundamental in determining the agronomic quality of water (BARROSO *et al.*, 2011).

However, in order to correctly interpret the quality of irrigation water, the parameters analyzed must be related to their effects on the soil, the crop and irrigation management, which will be necessary to control or compensate for problems related to water quality (BERNARDO *et al.*, 2006).

2.4 Soil and water salinities and their effects on plants

Soil salinization can be both natural and caused by anthropogenic action. However, in cultivated areas, it is often the result of human action through the adoption of agricultural practices that sometimes contribute to its worsening. Although saline soils can occur in the most diverse environmental conditions, this problem is more frequent in arid and semi-arid regions. This is mainly due to climatic conditions favorable to the concentration of salts on the soil surface, such as limited rainfall, high temperatures, high evaporative demand and irrigation practices that are not always appropriate, such as the use of water with a high saline concentration, or irrigation with incorrect management of the water blade (MELONI et al., 2003).

In many areas there are serious obstacles to the production system due to the process of soil salinization, even when irrigating with good quality water. The excess of dissolved salts in the soil solution, or in irrigation water, is one of the most serious abiotic problems in world agriculture as it causes declines in productivity and fruit size in a wide variety of cultivated plant species, including the melon tree. It is estimated that currently around 20% of cultivated land and approximately half of irrigated areas in the world are affected by salts; in Brazil, there are around 4.5 million salinized hectares, located mainly in the Northeast region, where the majority of irrigated perimeters are concentrated (MELO *et al.,* 2008).

The accumulation of soluble salts, in addition to reducing the osmotic potential of the soil solution, produces changes in pH, nutritional imbalance and disruption of its aggregates (MEDEIROS *et al.*, 2010). Tayer (1987) mentions three immediate effects of salinity on plants: physiological drought (A), caused by the decrease in osmotic potential; the toxic effect of ions (B), mainly chlorine and sodium; and nutritional imbalance (C) caused by the high ionic concentration, especially of sodium ions, inhibiting the absorption of other nutrients. The nutritional imbalance caused by salinity is due to excessive absorption of Na^+ and Cl^-, as well as disturbances in the absorption or distribution of various essential elements (AZEVEDO NETO and TABOSA, 2000; LACERDA, 2005).

Not all crops respond equally to salinity; some produce acceptable yields at high levels of salinity and others are sensitive to relatively low levels, the difference between which is due to their greater capacity for osmotic adaptation, which allows them to absorb a greater quantity of water, even in saline conditions (MEDEIROS *et al.*, 2011). According to the authors, plants that are more tolerant of saline environments increase the saline concentration inside them so that an osmotic gradient remains favorable for water absorption by the roots.

Another option to reduce the effect of water salinity on plants would be to use the crop phases in which there is greater tolerance and/or the mixture of saline water with low salinity water, since their use without deleterious effects depends on crop tolerance and irrigation management (FAGERIA *et al.*, 2011). Also, using a daily irrigation frequency, at any application period or divided between the morning and afternoon periods, resulted in better use of water resources and higher economic yields, which were found under the conditions of the melon crop study (CAMPELO *et at.*, 2014).

The excessive concentration of salts in the soil solution decreases the free energy of water, reducing its absorption by plants. The main effect of salts is osmotic, since the high level of salts in the soil makes it difficult for crops to absorb water, especially plants that are sensitive to salinity. Plants have a semi-permeable membrane that allows water to pass through, but prevents salts from passing through. The saltier the water, the more osmotically difficult it is to extract from the soil solution (BARROS, 2002).

Medeiros *et al.* (2007), studying water salinity levels in the melon crop, found that irrigation water from shallow wells is widely used. This water is of low quality, but has a low collection cost and a huge exploitable volume, and can be used as an alternative source by small, medium and large producers, expanding the area under melon cultivation. Results found by Medeiros *et al.* (2011) showed that an increase in water salinity has a negative influence on crop yield, with relative losses per unit increase in salinity above 0.54 dS m^{-1} , of 7.5%; 7.8%; 8.32% and 8.70%, for commercial and total productivity and number of commercial and total fruits, respectively, but the average mass of melon fruits was not affected.

Zong *et al.* (2011) also observed the effects of salinity on the production and quality of two Chinese cucurbits. For melon (cv. Huanghe), the authors recorded a decrease in yield and glutamic acid content as the salinity of the irrigation water increased. However, even if irrigation potentially has an effect on soil salinization, this does not imply that the soil is salinized to the point of being unsuitable for agriculture. There are numerous agricultural enterprises in the world where success has been achieved with crops irrigated with saline water, especially when there is constant monitoring of soil salinity and the adoption of saline water management practices with restricted use for irrigation, such as the research carried out by Kusvuran *et al.* (2011) and Zong *et al.* (2011), with the melon crop.

Medeiros *et al.* (2011) studied the yield and quality of cantaloupe melon grown with water of different salinity levels, and found that the level of salinity in the irrigation water negatively influenced the yield of the 'Sedna' melon hybrid and the loss of yield in response to salinity occurred through a reduction in the number of fruits per plant, while the average values of soluble solids and flesh firmness were not influenced by the salinity of the irrigation water.

According to Terceiro Neto *et al.* (2014), who studied the management strategies of irrigation with brackish water and found that it did not influence the accumulation of dry phytomass of the Pele de Sapo melon plants (cv Sancho). The increase in K absorption in the plant tissue of the Pele de Sapo melon (cv Sancho) contributes to maintaining the high K/Na ratio in the aerial part of the plants, proving the tolerance of this cultivar to salinity.

3 METHODOLOGY

3.1 Characterization of the experimental area

The experiment was carried out at the Federal University of Campina Grande at the Center for Agrifood Sciences and Technology (CCTA), located in the municipality of Pombal - PB during the period from July to October 2015 (Figure 1). The municipality of Pombal has the following geographical coordinates: 6° 46'13" south latitude and 37° 48'06" longitude west of Greenwich. The region's climate, according to the Koppen climate classification adapted to Brazil (COELHO & SONCIN, 1982), is of the BSh type, which represents a hot and dry semi-arid climate with average rainfall of 750 mm per year^{-1} and average annual evaporation of 2000 mm. The soil in the experimental area is of the Neossolo Fluvico type (EMBRAPA, 2008).

Figure 1 - Location of the experimental area with the melon crop. CCTA/UFCG, Pombal-PB, 2016.

3.2 Experimental design and treatments

The experiment was carried out in 2 x 4 subdivided plots using a randomized block design, with four replications. The plots consisted of 4 levels of irrigation water salinity (0.3, 1.0, 1.7 and 2.4 dS m^{-1}) and the sub-plots were made up of plants from seeds treated and not treated with saline NaCl solution, according to the treatments suggested in the study.

3.3 Preparing the soil in the experimental area

One week before the experiment was set up, soil samples were taken from a depth of 0-20 cm, dried in the air and sieved through a 2 mm mesh, after which they were sent to the Soil and Plant Nutrition Laboratory - LSNP, Campus Pombal-PB. The results of the soil analysis are shown in Table 1. The soil was then prepared by ploughing with a harrow, turning the soil to a depth of 0.20 m, raising the beds, setting up the drip irrigation system and fertilizing the foundations. The soil was covered using double-sided polyethylene film (black and silver) (Figure 2A).

Table 1. Chemical analysis of the soil in the experimental area, UFCG/CCTA/UAGRA. Pombal PB, 2016.

Variables	**Values obtained**
Potassium ($cmol_c$ dm $)^{-3}$	0,52
Sodium (cmolc dm $)^{-3}$	1,05
Calcium ($cmol_c$ dm $)^{-3}$	2,50
Magnesium (cmolc dm $)^{-3}$	2,00
Aluminum ($cmol_c$ dm $)^{-3}$	0,10
Hydrogen + Aluminum ($cmol_c$ dm $)^{-3}$	1,65
Phosphorus (mg dm $)^{-3}$	8,00
Base Sum - SB ($cmol_c$ dm $)^{-3}$	7,07
Cation Exchange Capacity - CEC ($cmol_c$ dm $)^{-3}$	8,72
Organic Matter (g kg $)^{-1}$	20,0
Sodium Saturation - PST %	1,36
pH in water (1:2.5)	6,65
Base Saturation - V %	81,07

P, K, NA: Mehlich 1 Extractor; Al, Ca, Mg: KCL 1M Extractor; SB = Ca^{+2} + Mg^{+2} + K^+ Na^+ ; H + Al: Calcium Acetate 0.5 M Extractor; pH 7.0; CTC = SB + H^+ + Al^{+3} ; M.O.: Walkley-Black Wet Digestion; PST = Percentage of Exchangeable Sodium, according to Embrapa methodology (1997).

Foundation fertilization was managed in accordance with the soil analysis and the recommendations for melon growing, applying the following fertilizers: P O_{25} (160 kg ha^{-1} of simple superphosphate), proportional to 100% of the foundation fertilization, together with 10% of the N and K2O of the recommended dose, equivalent to 17 and 10 kg ha^{-1} in the form of urea and potassium chloride, respectively.

For top dressing, N and K2O were used at doses of 153 and 90 kg ha^{-1} , corresponding to the remaining 90%. They were applied via fertigation every week for eight weeks, starting three days after transplanting. Each week they were applied, respectively, as shown in Table 2.

Table 2. Percentage values of top dressing fertilization with N and K2O, in the experiment, UFCG/CCTA/UAGRA. Pombal PB, 2016.

WEEKS								
FERTILIZERS *(%)*	**1ª**	**2ª**	**3ª**	**4ª**	**5ª**	**6ª**	**7ª**	**8ª**
NITROGEN (N)	**5**	**8**	**10**	**12**	**15**	**20**	**12**	**8**
POTASSIUM CHLORIDE (K2O)	**5**	**8**	**10**	**12**	**15**	**20**	**12**	**8**

Agrotextile was used to cover the entire area of the plants and was supported by the recycled rigid polyethylene hoops, at a height of 40 cm from the ground in the central part of the hoop. For the experiment, a white polypropylene agrotextile was used, with a width of 1.38 m and a weight of 15 g cm^2 (Cultivar Agricola Comércio Importadora e Exportadora) and placed and fixed on the sides forming a tunnel (Figure 2B).

Figure 2 - Mulching and agrotextile in the experimental area. CCTA/UFCG, Pombal-PB, 2016.

3.4 Planting the crop

The melon cultivar used was 'Rangers' from the Cantaloupe group. It was planted at a spacing of 2.0 x 0.4 m, with one plant per hole.

The seeds were soaked in saline solution (NaCl) and water without affecting their germination (Figure 3A). On July 6, 2015, the NaCl and H_2O solutions used, at electrical conductivity levels of 9.00 and 0.3 dS m^{-1} , respectively, were applied to the seeds and their soaking time, after selection, was 10 hours, according to Queiroga *et al.* (2006) and Pereira *et al* (2012), and these were the best results found with the pre-test of melon seeds when the seeds were soaked for 10 hours and the solution

chosen was 9.73 dS m .$^{-1}$

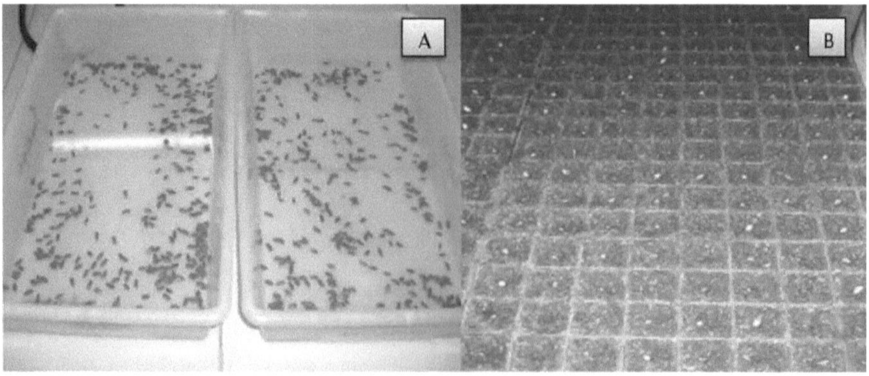

Figure 3 - Preparation of seeds in NaCl solution and sowing in trays. CCTA/UFCG, Pombal-PB, 2016.

Sowing took place on July 6, 2015 (Figure 3B), in 128-cell polyethylene trays, the substrate used was Tropstrato HA vegetables; subsequently, each tray was sown with one seed per cell (treated or not) in NaCl saline solution.

The seedlings were transplanted fifteen days after sowing (on July 21, 2015), when they already had their second definitive leaf.

Irrigation was carried out using an aqueous solution with different salinity levels, obtained by adding NaCl to the irrigation water (saline concentration levels: S1 = 0.3; S2 = 1.0; S3 = 1.7 and S4 = 2.4 dS m^{-1}), and monitored using a portable conductivity meter (Sammar CD-840). The irrigation rates were applied according to the crop's stage of development.

The plants were managed according to the suggested treatments; the agrotextile was removed on August 12, 2015 when the plants were 22 days after transplanting (DAT) and at the beginning of the crop's flowering so that bees could pollinate them.

When the agrotextile was removed, manual weeding was carried out in each treatment and plant health was controlled with a fungicide and insecticide registered for the crop (Conect 750g⁄400L and Kocide 750 g/ha). Three applications were made, one when the agrotextile was removed, the second 15 days after the first application and the last eight days before the harvest due to the occurrence of whitefly (*Bemisia argentifolii*) in the final phase of the experiment. The products were applied in the late afternoon, avoiding the application of any type of insecticide in the morning when there was a greater presence of bees.

The fruit was harvested on September 22nd and 23rd. The fruit was picked when the peduncle was cracked and uniformly lacy, which is a reliable indication of this cultivar's harvest point. The crop cycle lasted 78 days, from sowing to the start of the harvest (Figure 4).

Figure 4 - Melon harvest in the experimental area. CCTA/UFCG, Pombal-PB, 2016.

3.5 Characteristics evaluated

3.5.1 Evaluations of the components of crop yield formation

The number of fruits per plant was obtained by counting the fruits per plot in each treatment and on all useful plants.

The fruit mass was obtained from the ratio between the production of the plot and the number of fruits in the plot, in kg fruit $.^{-1}$

The total melon yield was calculated by adding up all the harvests in the plot and then estimating them for 1 ha, expressed in Mg ha $.^{-1}$

3.5.2 Physical characteristics of the fruit

A sample of three fruits was taken from each experimental unit and each melon fruit was cut lengthways in order to obtain the following characteristics: flesh firmness, using a McCormick model FT 327 manual penetrometer with a 12 mm diameter cylindrical tip; longitudinal and transverse diameters of the fruit using a ruler graduated in centimeters (cm); fruit shape index, obtained from the ratio of the longitudinal diameter to the transverse diameter of the fruit; mesocarp thickness, obtained by taking two readings, one in each half of the fruit, of the internal cavity of the melon fruit, this being the flesh of the fruit, taking the readings in the equatorial region of the fruit, using a digital caliper (cm).

3.5.3 Fruit chemical characteristics

Soluble solids (%) and titratable acidity were evaluated from the same sample of three fruits, in samples of fruit slices taken lengthwise and homogenized in a fruit centrifuge to obtain the juice; the soluble solids (TSS) were determined using a digital refractometer, obtaining the values in percentage. For titratable acidity (TA) (percentage of citric acid), a 2 mL aliquot of juice was used in duplicate, to which 50 mL of distilled water and two drops of 1% alcoholic phenolphthalein were added, followed by titration with 0.1 N NaOH solution until the turning point; with the same sample, the ripeness index was determined using the ratio between soluble solids and titratable acidity.

3.6 Data analysis

The data obtained was subjected to analysis of variance, F test, using SAEG 9.0 software. For the qualitative treatment (seed soaking in NaCl) Tukey's test was carried out at 5% probability level and for the quantitative treatment (water salinity levels) regression analysis was carried out using Table Curve 2D software.

4 RESULTS AND DISCUSSION

4.1 Evaluations of the components of crop yield formation

There was a significant interaction between the factors seed treatment with NaCl and water salinity levels on the number of fruits per plant and yield, while for fruit mass there was a significant effect only of seed treatment with NaCl ($p < 0.05$) (Appendix A).

For fruit mass, the highest value was found, an average of 1.00 kg $fruit^{-1}$, when the plants came from seeds treated with NaCl (Table 3). In this sense, the prior action of osmoconditioning may have favored the plants' tolerance to adverse conditions, i.e. regardless of the level of salinity applied. Tolerance to stress is one of the main advantages of osmoconditioning, as it gives the plants greater tolerance to a sharp drop or rise in temperature, water deficiency and an increase in saline concentration, which is often seen in the literature. Examples of this are the results obtained by Aguiar (1979), showing benefits to the performance of rice seeds under saline stress, and by Eira & Marcos Filho (1990), in lettuce seeds, the effects of osmotic conditioning with PEG and mannitol.

In terms of the number of fruits per plant, the lowest salinity level of 0.3 dS m^{-1} showed significant values in plants from seeds not treated with NaCl. Under more extreme conditions of irrigation water salinity, as the level increased to 2.4 dS m^{-1}, it became clear that the effect of NaCl-treated seeds was equal to that of seeds not treated with NaCl.

The total yield showed a different response, probably because the mass of the fruit played a greater role in shaping the crop's yield than the number of fruits per plant.

According to Bruce *et al.* (2007), an initial stress caused by physiological conditioning favors a tolerance response to a future stress that the plant may suffer. According to the same authors, plants are capable of expressing a type of "memory", also called a "stress imprint", and this imprint is commonly translated into genetic and biochemical modifications induced by a first exposure to stress that increase resistance to a subsequent adverse condition.

Table 3 - Average values of fruit mass, number of fruits per plant and total yield of melon as a function of seed treatment with and without NaCl. CCTA/UFCG, Pombal - PB, 2015.

Seed treatment	Fruit mass (kg fruit)-1	Number of fruits per plant				Total Productivity (Mg.ha)-1			
		Salinity level (d m-1)				Salinity level (dS m)-1			
		0,3	1,0	1,7	2,4	0,3	1,0	1,7	2,4
In the presence of NaCl	1,006 a	2,17 b	2,48 a	2,35 a	2,25 a	36,70 a	32,30 a	29,49 a	25,27 a
In the absence of of NaCl	0,895 b	2,82 a	2,56 a	2,36 a	2,27 a	24,98 b	28,51 a	26,55 a	23,78 a
CV (%)	9,67			6,52				10,33	

*Means followed by the same letter in the columns do not differ at the 5% probability level using the Tukey test.

It was found that, regardless of seed treatment with NaCl, there was a reduction in fruit mass as the water salinity level increased to 2.4 dS m- 1. The observed reduction in fruit mass was 6.7% (Figure 6). Thus, the plants in the field showed that as the adverse salinity conditions to which they were subjected increased, the fruit tended to decrease in mass, restricting the development of fruit weight, which indicates that this melon cultivar has tolerance to salinity.

It should also be noted that the negative effect of salinity on fruit mass was probably due to the difficulty plants had in absorbing water, as well as the entry of ions in toxic concentrations, which reduced water absorption by the plants and consequently altered the process of increasing fruit mass. This factor did not affect the crop's productivity. According to Pacheco *et al.* (2012) salinity reduces the osmotic potential of the medium and prolongs the time needed for plants to absorb water.

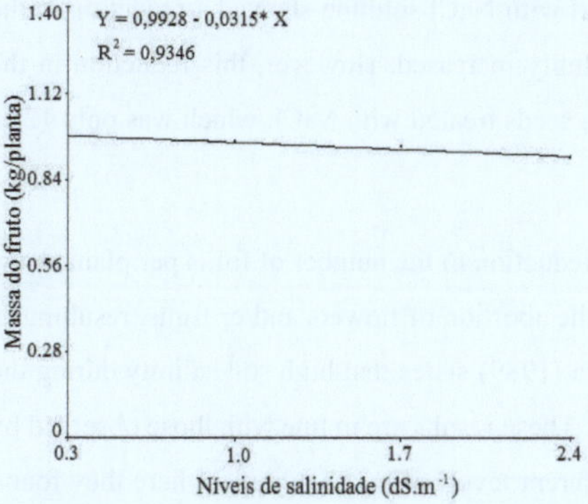

Figura 5 - Adjusted response function for melon fruit mass as a function of water salinity levels. CCTA/UFCG, Pombal-PB, 2016.

Silva et al. (2014), the production of zucchini under different levels of saline water, resulted in a higher average mass of total fruit in the conductivity of 1.5 dS m^{-1} followed by a decrease with the continuous increase in the electrical conductivity of the irrigation water.

In seeds treated with NaCl, the maximum value for the number of fruits per plant was 2.48 at a water salinity level of 0.90 dS m^{-1} , with a subsequent reduction in the number of fruits per plant as the water salinity level increased (Figure 6A). In this case, the reduction was 12% in the number of fruits. On the other hand, the plants whose seeds were not treated with NaCl showed a decreasing linear response as the water salinity level increased up to 2.4 dS m^{-1} (Figure 6B). In this case, increasing the electrical conductivity of the water led to a 25.9% reduction in the number of fruits per plant.

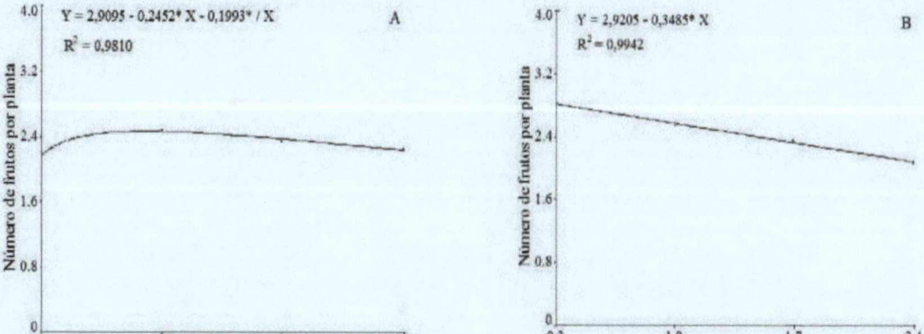

Figura 6 - Adjusted response function for the number of fruits per melon plant in plants from seeds treated (A) and not treated (B) with NaCl. CCTA/UFCG, Pombal-PB, 2016.

As described above, both seeds treated and untreated with NaCl solution showed a reduction in the number of fruits per plant as the level of water salinity increased. However, this reduction in the number of fruits on the plant was less in plants from seeds treated with NaCl, which was only 12%, which may be due to osmoconditioning.

One of the factors that may have contributed to the reduction in the number of fruits per plant as the level of salinity of the irrigation water increased is the abortion of flowers and/or fruits, resulting in a reduction in the number of fruits per plant. Cuenca (1989) states that high soil salinity during the flowering period can reduce the number of fruits set. These results are in line with those observed by Silva *et al.* (2005), working with melon trees at different levels of water salinity, where they found that the number of total fruits on the 'Trusty' hybrid decreased linearly, obtaining percentage differences of 13% and 26% at salinities of 2.5 dS m^{-1} and 4.4 dS m^{-1} , respectively.

Terceiro Neto *et al.* (2013) found a reduction in the number of commercial (NFC) and total (NFT) fruits as the salinity of the irrigation water increased, while Medeiros *et al.* (2011) found that the number of total fruits per plant decreased by 8.7% as the salinity of the irrigation water increased.

For total productivity, at a water conductivity of 0.3 dS m^{-1} , there was a tendency for plants from NaCl-treated seeds to yield more when compared to untreated seeds (Table 3). This response was different from that observed in the number of fruits per plant, probably due to the fact that fruit mass contributes more to crop productivity than the number of fruits on the plant. Another fact is that the fruit from plants treated with NaCl during seed soaking had a higher mass and, consequently, a lower number on the plant due to the source-drain balance in the plant.

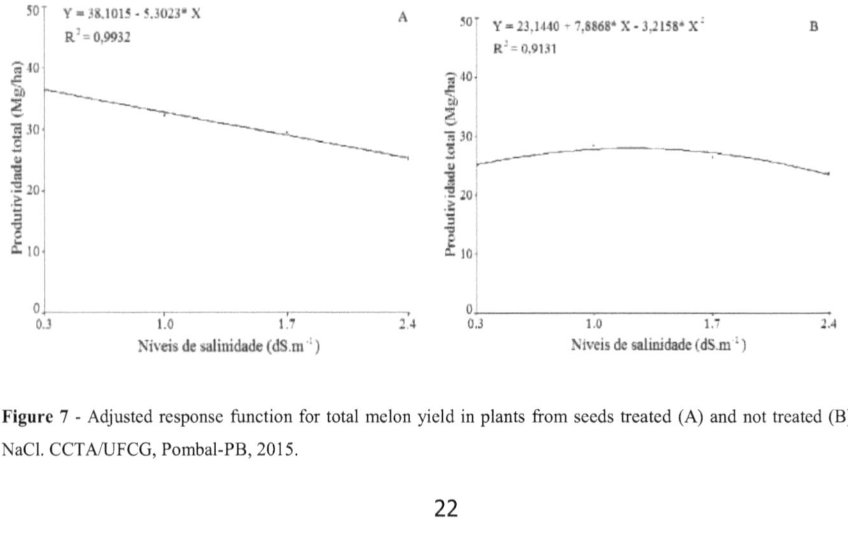

Figure 7 - Adjusted response function for total melon yield in plants from seeds treated (A) and not treated (B) with NaCl. CCTA/UFCG, Pombal-PB, 2015.

Considering the salinity level of the irrigation water, there was a decreasing linear response as the salinity level increased up to 2.4 dS m^{-1} with a maximum estimated value of 36.51 Mg ha^{-1} at the 0,3 dS m^{-1} (Figure 7A) and a quadratic response with a maximum estimated value of 27.97 Mg ha^{-1} at a water salinity level of 1.23 dS m^{-1} (Figure 7B), in plants from seeds treated or not with NaCl.

According to the results of the experiment, the main cause of the reduction in total productivity at higher levels of water salinity is due to a reduction in the number of fruits per plant and fruit mass. Mascarenhas *et al.* (2010) found similar results when they observed that when the salinity of irrigation water was increased in the production of the Gàlia melon hybrid 'Néctar', there was a reduction in commercial productivity of 10.9 Mg ha^{-1} for water with a salinity of 3.9 dS m^{-1} compared to 0.54 dS m^^{1} , representing a relative loss of 35.3%.

Freitas *et al.* (2014) found similar results for the total yield of melons subjected to different levels of salinity when they verified that there was a significant fit of the linear model and that the increase in salinity decreased the values of the yield components linearly. The decrease in productivity observed by these authors was 4.24 Mg ha^{-1} per unit increase in the electrical conductivity of the irrigation water for total production.

In agreement with Medeiros *et al.* (2011), the total productivity of the melon hybrid 'Medellin' was negatively influenced by increasing salinity levels, with the highest value being obtained at 0.54 dS m^{-1} of 57.73 t ha^{-1} . The unit increase in salinity above the lowest level (0.54 dS m^{-1}) resulted in a 7.8% reduction in total commercial production. This study found a different result in that the main cause of the reduction in fruit yield was the decrease in the number of commercial fruits per plant when the levels of salts in the water were high, since the reduction in mass was not significant.

$^{-1}$According to Medeiros *et al.* (2008), the salinity of the irrigation water had a significant effect on the total productivity of the Orange Flesh and Trusty melon cultivars, which achieved a production of 52.2 M ha^{-1} with the lowest salinity irrigation water (ECa = 1.1 dS m^{-1}), which was reduced to 36.48 Mg ha^{-1} with the most saline water (ECa = 4.5 dS m).

In relation to the total yield variable in plants from seeds not treated with NaCl (Figure 7B), there was a significant quadratic effect, with increasing salinity of the irrigation water affecting total fruit yield. Initially, there was a positive effect up to a salinity of 1.22 dS m^{-1} (27.98 M ha^{-1}) and a reduction from this level onwards, so that the lowest value was obtained at the highest salinity (2.4 dS m^{-1}) with 23.78 Mg ha^{-1} , which meant a 15.83% decrease in productivity as salinity increased. This suggests that higher levels of salinity produce less fruit, negatively affecting the final yield of the crop.

Reduction in melon productivity has been a common behavior when irrigation water with a high concentration of salts is used. The study conducted by Kusvuran *et al.* (2011) with 31 melon genotypes showed great genotypic variation in the melon tree in response to the effects of salinity. According to Botia *et al.* (2005), this genetic variability is associated with the mechanisms of adaptation to salt stress developed by genotypes of salt-tolerant species and, in the case of the melon tree, the authors report that melon plants, when tolerant to salinity, accumulate Na^{+} and Cl^{-} in the stem, avoiding high concentrations of these ions in the leaves and, consequently, toxicity.

4.2 Physical characteristics of the fruit

There were no significant differences in the analyzed characteristics of length (cm), diameter (cm), shape index, firmness (N) and thickness (cm) of the melon fruit pulp depending on whether or not NaCl was used in the seed treatment (Table 4). The treatment of the seeds with NaCl did not have a positive effect on the physical characteristics of the fruit, although there was a tendency, in absolute terms, for the fruit from plants grown from seeds treated with NaCl to have higher values for length, diameter and thickness of the flesh, i.e. variables that involve the size of the fruit.

Table 4 - Average values of length, diameter, fruit shape index, firmness and pulp thickness of melon fruit as a function of seed treatment with and without NaCl. CCTA/UFCG, Pombal - PB, 2016.

Seed treatment	**Length (cm)**	**Diameter (cm)**	**format index**	**Firmness (N)**	**Pulp thickness (cm)**
In the presence of NaCl	15,09 a	13,83 a	1,09 a	25,53 a	4,12 a
In the absence of NaCl	15,04 a	13,72 a	1,09 a	24,65 a	4,07 a
CV (%)	4,14	3,53	2,17	9,05	3,43

*Means followed by the same letter in the columns do not differ at the 5% probability level using the Tukey test.

A quadratic response was observed for the length and diameter of the fruit with maximum estimated values of 16.2 and 14.3 cm, obtained at water salinity levels of 0.48 and 1.37 dS m^{-1} (Figure 9A and B). From this level there was a reduction in fruit length and diameter of 10.0 and 3.3% compared to the salinity level of 2.4 dS m .$^{-1}$

The fruit shape index (Figure 9C) showed a decreasing linear response with an estimated reduction of 4.5% with increasing water salinity levels up to 2.4 dS m^{-1} . Therefore, the values found for length and diameter were not enough to significantly alter the fruit shape index, which showed values close to 1 (one). This gives the fruit a spherical shape. In melons, the shape index is an important quality attribute in classification and standardization, and can determine the acceptance and value of the product for certain markets. It also defines the packaging and arrangement of the fruit inside. Therefore, fruits with a shape index close to 1.0 are preferred, since above (elongated) and below (flattened) this value compromises their accommodation in packaging (PURQUERIO & CECILIO FILHO, 2005).

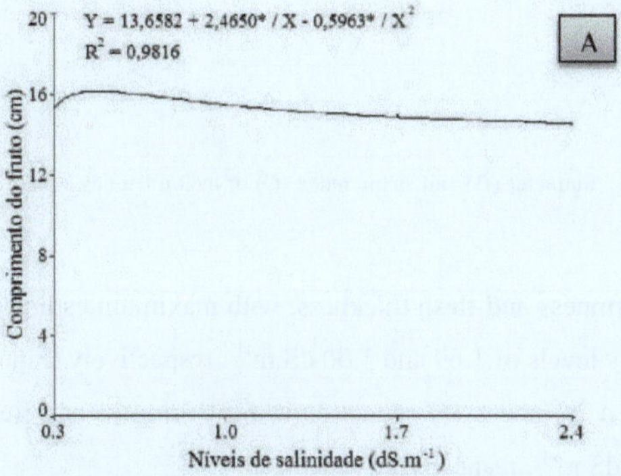

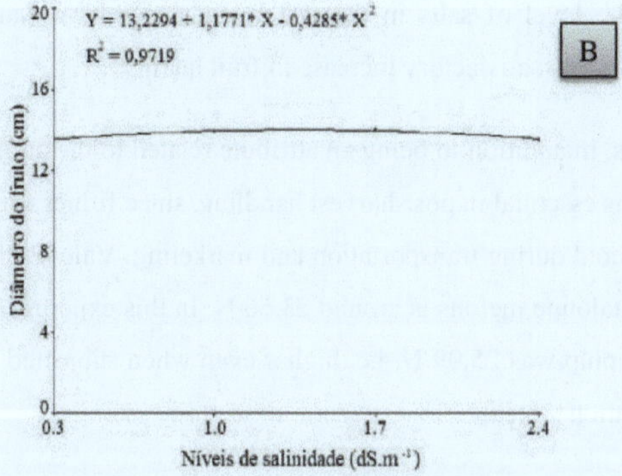

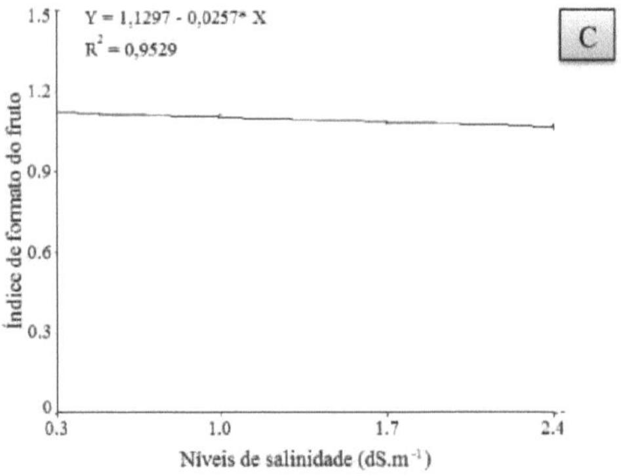

Figure 8 - Adjusted response function for the length (A), diameter (B) and shape index (C) of melon fruit as a function of water salinity level. CCTA/UFCG, Pombal-PB, 2016.

Quadratic responses were recorded for fruit firmness and flesh thickness, with maximum estimated values of 25.9 N and 4.2 cm obtained at salinity levels of 1.69 and 1.00 dS m^{-1} , respectively (Figure 9A and B). From this salinity level there was a 2.1 and 5.3% reduction in fruit firmness and flesh thickness compared to the salinity level of 2.4 dS m^{-1} , respectively.

The effect of osmotic pressure caused by the level of salts in the water tends to inhibit water absorption by the plant, favoring flesh firmness as a satisfactory increase in fruit quality.

Thomaz *et al.* (2009) stated that pulp firmness, in addition to being an attribute related to the aroma and flavor of the fruit, is important because it is essential in post-harvest handling, since firmer fruits are more resistant to mechanical injuries suffered during transportation and marketing. Vale (2000) explains that the firmness of the pulp for Cantaloupe melons is around 23.56 N. In this experiment, the average value of the firmness of the fruit pulp was 25.09 N, i.e. higher even when subjected to osmoconditioning and the salinity of the irrigation water.

Mascarenhas *et al.* (2010), working on the production and quality of Galia melons grown under different salinity levels, analyzed the firmness of the pulp and concluded that although they observed a homogeneous behavior of the data with increasing salinity, the treatment 3.9 dS m^{-1} showed the highest average value for this characteristic (38.0 N).

The thickness of the pulp is a characteristic that allows us to estimate and obtain its yield, demonstrating greater use of the edible part. Thus, it can be said that the fruit from treatments with seeds from osmoconditioning and then subjected to different levels of salinity did not differ greatly from each other. In this sense, the thickness of the flesh follows the growth of the fruit in terms of length and diameter, showing a response proportional to its growth.

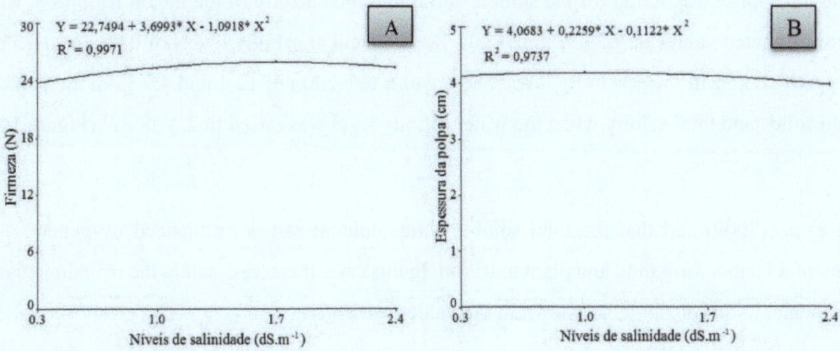

Figure 9 - Response function adjusted for firmness (A) and pulp thickness (B) of melon fruit as a function of water salinity level. CCTA/UFCG, Pombal-PB, 2016.

4.3 Chemical characteristics of the fruit

Osmoconditioning the seed with NaCl did not significantly alter the soluble solids content, total acidity and ripeness index of the fruit (Appendix C). The soluble solids content is expressed as a percentage of the mass of fresh material, which in melons is one of the indicators of quality and the ideal point of harvest, since sugars are the main soluble components measured when estimating soluble solids.

Regardless of the cultivar, the soluble solids content found in the literature must be at least 9^0 Brix to obtain an acceptable taste. In this study, the averages obtained in the treatments were below this standard, with the highest value found being 8.88% in the treatment where the seeds were treated with NaCl (Table 5).

Table 5 - Average values of soluble solids (%), total acidity (%) and ripeness index of melon fruit as a function of seed treatment with and without NaCl. CCTA/UFCG, Pombal - PB, 2016.

Treatment of seed	Soluble solids (%)	Total acidity (% of citric acid)	Index of fruit ripening
In the presence of NaCl	8,88 a	0,090 a	110,96 a
In the absence of NaCl	8,63 a	0,105 a	92,03 a
CV (%)	11,29	23,18	20,69

*Means followed by the same letter in the columns do not differ at the 5% probability level using the Tukey test.

A quadratic response was found for the soluble solids and total acidity of the melon fruit pulp, with maximum estimated values of 9.59 % and 0.107 % citric acid at salinity levels of 0.49 and 1.75 dS m^{-1} , respectively. From these salinity levels there was a reduction of 12.1 and 4.7 % in the content of soluble solids and total acidity when the water salinity level was raised to 2.4 dS m^{-1} (Figure 10A and B).

Siqueira *et al.* (2009) add that the total soluble solids content can be influenced by genetic and environmental factors, irrigation and plant nutrition. In this case, the excess salt in the irrigation water was detrimental to this attribute, as shown in the values presented.

A different result was found by Mascarenhas *et al.* (2010), working with melon, in which the soluble solids values increased with the increase in the salinity levels of the irrigation water, with a value of 9.08% for salinity 3.9 dS m^{-1} . Medeiros *et al.* (2008), studying three levels of irrigation water salinity in the field, observed an increasing linear effect on the fruit soluble solids of the melon tree as a result of the increase in salinity. Costa *et al.* (2013), working with watermelon, found that the effect of salinity influenced the soluble solids content of the 'Shadow' cultivar, in which the increase in the salinity levels of the irrigation water from 2.77 to 4.91 dS m^{-1} raised the soluble solids values from 3.58 to 5.08%, respectively.

As for total acidity, the variation in acidity levels in melons has little significance due to its low concentration, and the impact of acidity on flavor is not very significant (MORAIS *et al.,* 2009).

According to Silva (2004), the loss of acidity is considered desirable in most fruit and important for the ripening process, where it is probably converted into sugars. Albuquerque *et al.* (2006) state that organic acids, together with sugars, enhance the perception of the specific flavor of melons.

The titratable acidity of a fruit is given by the presence of organic acids which decrease as ripening progresses due to oxidation in the tricarboxylic acid cycle, being fundamental in the synthesis of phenolic compounds, lipids and volatile aromas (CHITARRA & CHITARRA, 2005).

The ripeness index showed a decreasing linear response with a maximum estimated value of 113.7 at a salinity level of 0.3 dS m^{-1} (Figure 10C). As the salinity level of the water rose to 2.4 dS m^{-1} , there was an observed reduction of 21.5% in the value of this ratio between soluble solids and total acidity of the melon fruit pulp.

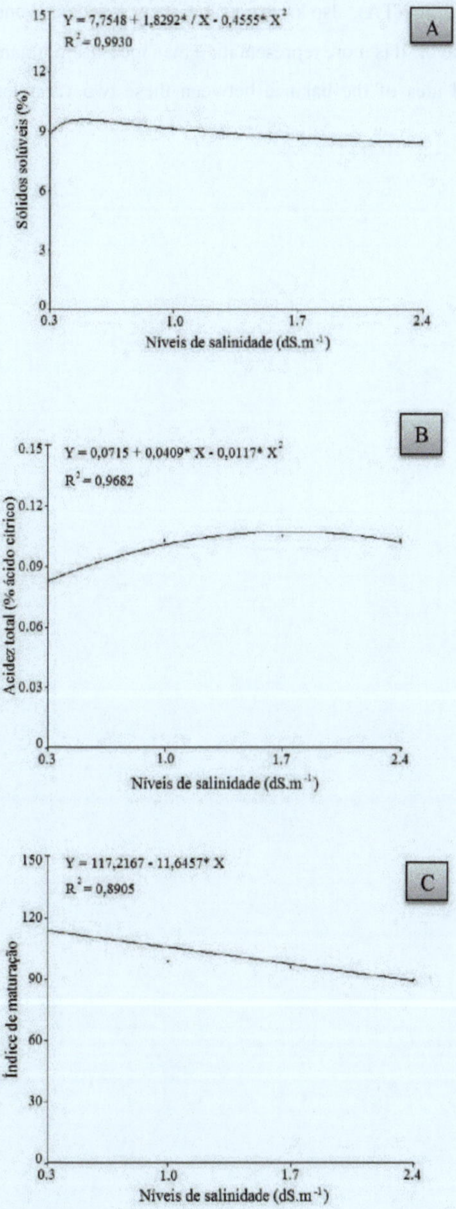

Figure 10 - Adjusted response function for soluble solids (A), total acidity (B) and ripeness index (C) of melon fruit pulp as a function of water salinity level. CCTA/UFCG, Pombal-PB, 2016.

The ratio between soluble solids (SS) and total acidity (TA), also known as the ripeness index, is one of the most widely used ways of assessing fruit flavor. It is more representative than measuring sugars or titratable acidity alone and provides a good idea of the balance between these two variables (CHITARRA & CHITARRA, 2005).

5 CONCLUSIONS

Osmoconditioning the seeds in NaCl solution affected the production components of the melon tree.

Fruit mass was higher in plants from seeds treated with NaCl.

Fruit mass contributed more than the number of fruits per plant to crop yield, which was only higher under low water salinity conditions of 0.3 $dS_{m\ -1}$.

The quality of the melon fruit was not altered by osmoconditioning the seed with NaCl.

Plants irrigated with water with a salinity of 2.4 dS $_{m-1}$ reduced their melon fruit yield and quality.

6 BIBLIOGRAPHICAL REFERENCES

AYERS, R. S.; WESTCOT, D. W. 1999. **Water quality in agriculture**. Translated by GHEYI, H. R.; MEDEIROS, J. F.; DAMACENO, F. A. V. Campina Grande: UFPB, 153 p. (Estudos FAO 29, 1999).

ALBUQUERQUE et al. 2006. Crep evapotranspiration Guidelines for computing crep water requirements. Rome: **FAO**, 279 p. (FAO, Irrigation and Drainage Paper, 56).

AGUIAR, P. A. A. 2009. Pre-treatment of rice seeds as a means of overcoming the effect of salinity on germination and vigor. **Revista Brasileira de Sementes**, Brasilia, v. 1, n. 1, p. 65-70.

AGRIANUAL 2012 - **ANUARIO DA AGRICULTURA BRASILEIRA**. Sao Paulo: FNP. Sao Paulo, October 2011. 512p.

ALVES, S.S.V. **Performance of crops under saline stress in representative soils of the Mossoró-Assu agropolis**. 2012. Thesis (Doctorate). Federal Rural University of the Semi-Arid Region. MOSSORÓ - RN.

AMOOAGHAIE, R.; NIKZAD, K.; SHAREGHI, B. 2010. The effect of priming on emergence and biochemical changes of tomato seeds under suboptimal temperatures. **Seed Science and Technology**, Zurich, v. 38, n. 2, p. 508-512.

ANDRADE JÙNIOR, W.P.; PEREIRA, F.H.F.; FERNANDES, O.B.; QUEIROGA, R.C. F.; QUEIROGA, F.M. 2011. Effect of potassium nitrate in reducing salt stress in melon trees. **Revista Caatinga**, vol. 24, no. 3. p. 110-119.

ARAÙJO, J.L. P.; VILELA, N.J. **Socioeconomic Aspects**. In: SILVA, H.R.; COSTA, N. D. 2002. Melon, Production and Technical Aspects. Brasilia: Embrapa, Chap. 2, p. 15-18. (Frutas do Brasil, 33).

ASGHARIPOUR, M.R.; RAFIEI, M. 2011. The effects of osmo-priming on tomato seed germination. **Advances in Environmental Biology**, Amman, v. 5. N. 9, p. 2866-2869.

AZEVEDO NETO, A.D.; TABOSA, J.N. 2000. Salt stress in maize seedlings: Part II: distribution of cationic macronutrients and their relationship with sodium. **Revista Brasileira de Engenharia Agricola e Ambiental**, 4: 25 - 32.

BARROS, A.D. **Management of drip irrigation with different levels of water salinity in the**

melon crop. 2002. (Doctoral Thesis). Faculty of Agronomic Sciences, 124p.

BARROSO, A.A.F.; GOMES, G.E.; LIMA, A.E.O.; PALACIO, H.A.Q.; LIMA, C.A. 2011. Evaluation of water quality for irrigation in the South Central region of the State of Cearâ.

Revista Brasileira de Engenharia Agricola e Ambiental, v.15, n. 6.

BERNARDO, S.; SOARES, A.A.; MANTOVANI, E.C. 2006. **Irrigation manual**. 8 ed.

Viçosa: Ed. UFV. p. 611.

BESSA A.T.M.; LOPES W.A.R.; SANTOS F.G.B.; NEGREIROS M.Z.; MEDEIROS J.F.; SOARES A.M.; HONORATO A.R.F. 2011. Growth of 'Caribbean Gold' melon grown under protection with white agrotextile at different times of plant development. **Horticultura Brasileira**: v. 29, n. 2.

BIANCO, V.V. & PRATT, H.K. 1977. Compositional changes in muskmelon during development and in response to ethylene treatment. Journal of the American Society for **Horticultural Science**, 102, 127-133

BIRUEL, R.P.; BORBA FILHO, A.B.; ARAÙJO, E.C.E.; FRACCARO, F.O.; PEREZ, S.C.J.G.A.. 2007. Effects of conditioning followed or not by drying on Pterogyne nitens TUL. seeds under stress. **Ciência Florestal,** 17: 119-128.

BITTENCOURT, M.L.C.; DIAS, D.C.F.S.; DIAS, L.A.S.; ARAÙJO, E.F.. 2004. Effect of osmotic conditioning of seeds on germination and growth of asparagus seedlings. **Revista Brasileira de Sementes**, v.26, n.1, p.50-56.

BOTiA, P.; NAVARRO, J.M.; MARTINEZ, V. 2005. Yield and fruit quality of two melon cultivars irrigated with saline water at different stages of development. **European Journal of Agronomy**, 23: 243-253.

BRUCE, T.J.A.; MATTHES, M.C.; NAPIER, J.A.; PICKETT, J.A. 2007. Stressful "memories" of plants: evidence and possible mechanisms. **Plant Science**, v.173, p.603-608. Available at: <http://www.sciencedirect.com/science/article/pii/S0168945207002476>. Accessed on: July 5, 2016.

CAMPELO, A.R.; AZEVEDO, B.M.; NASCIMENTO NETO, J.R.; VIANA, T.V.A.; PINHEIRO NETO, L.G.; LIMA, R.H. 2014. Management of the melon crop subjected to irrigation frequencies and nitrogen fertigation. **Horticultura Brasileira**, v: 32, n. 2.

CHEN K., ARORA R. 2013. Priming memory invokes seed stress-tolerance. **Environ**. Exp. Bot. 94 33-45. 10.1016/j.jprot.2013.08.010

CHITARRA, M.I.F.; CHITARRA, A. B. 2005. Post-harvest of Fruits and Vegetables: Physiology and handling. 2 ed. **Revised and expanded.** Lavras: UFLA, 785p.

COSTA, A.R.F.C.; MEDEIROS, J.F.; PORTO FILHO, F.Q.; SILVA, J.S.; COSTA, F.G.B.; FREITAS, D.C.. 2013. Production and quality of watermelon grown with water of different salinities and doses of nitrogen. **Revista Brasileira Engenharia Agricola Ambiental**, v.17, n.9, p.947-954.

COSTA, N.D.; GRANGEIRO, L.C. 2003. **Fruit chemical composition and uses**. In: SILVA, H. R. da; COSTA, N.D. (eds). MELÂO: Production and Technical Aspects. Brasilia: Embrapa Communication for Technology Transfer, chap.4, p 22.

CUENCA, R.H. 1989. **Irrigation system design: an engineering approach**. Englewood Cliffs: Prentice Hall, 552p.

DIAS, R.C.S. et al. 1998. ed. **Production chains and natural systems: technological prospecting**. Brasilia: Embrapa-SPI; Embrapa-DPD. Chap. 17, p.441-494. Available at: <http://www.scielo.br/pdf/cagro/v33n5/v33n5a13.pdf>.

EIRA, M.T.S.; MARCOS FILHO, J. 1990. Osmotic conditioning of lettuce seeds: effects on germination. **Revista Brasileira de Sementes**, Brasilia, v. 12, n. 1, p. 9-27.

ELMSTROM, G.W.; MAYNARD, D.N. 1992. Exotic melons for commercial production in humid regions. **Acta Horticulturae**, 2: 117-123.

EMBRAPA. National Soil Research Center. 2008. **Brazilian Soil Classification System**. Brasilia: Embrapa-SPI; Rio de Janeiro: Embrapa-Solos, 306 p.

FAGERIA, N.K., BALIGAR, V.C., JONES, C.A. 2011. **Nutrient flux in soil-plant system: Growth and mineral nutrition in wheat crops,** third ed. CRC Press, pp. 57-79.

FIGUEIRÊDO, Vladimir Batista. 2008. **Evapotranspiration, growth and production of watermelon and melon irrigated with water of different salinities**: Vladimir Batista Figueirêdo. - 104 f. Thesis (Doctorate) - Universidade Estadual Paulista, Faculdade de Ciências Agronômicas de Botucatu.

FONTES, P.C.R.; PUIATTI, M. 2005. **Melon cultivation**. In: FONTES, P.C.R. (ed). Olericultura: teoria e pràtica. Viçosa: UFV, Cap.26. p. 407-428.

FREITAS, L.D.A.; FIGUEIRÊDO, V.B.; PORTO FILHO, F.Q.; COSTA, J.C.; CUNHA, E.M. 2014. Growth and production of melon cultivated under different levels of salinity and nitrogen. **Revista Brasileira Engenharia Agricola Ambiental**, v.18, (Suplemento), p. S20-S26.

GOMES, D.P.; SILVA, A.F.; DIAS, D.C.F.S.; ALVARENGA, E.M.; PANOZZO, L.E.; SILVA, L.J. 2012. Priming and drying on the physiological quality of eggplant seeds. **Horticultura Brasileira.** Vitoria da Conquista, v. 30, n. 3, p.484-488.

GRANGEIRO, L.C. et al. 2002. **Yellow melon - cultivation**. Jaboticabal, UNESP. 30 p.

GURGEL, M.T.; OLIVEIRA, F.H.T.; HANS, R.G.; FERNANDES, P.D.; UYEDA, C.A. 2010. Post-harvest quality of melon varieties produced under salt stress and potassium doses. **Revista Brasileira Ciência Agrària**. Recife, v.5, n.3, p.398-405.

HEYDECKER, W.; HIGGINS, J.; GULLIVER, R.L. 1973. Accelerated germination by osmotic seed treatment. **Nature**, v. 246, p. 42-44.

BRAZILIAN INSTITUTE OF GEOGRAPHY AND STATISTICS (IBGE). 2014. **Municipal agricultural production: temporary and permanent crops**. 99p.

KUsVURAN, s.; DAsGAN, H.Y.; ABAK, K.; AYDoNER, G. 2011. Determination of genotypical differences of melons to salt tolerance. **Acta Horticulturae**, 918: 777-783.

LACERDA, C.F. Interaction between salinity and mineral nutrition. In: NOGUEIRA, R.J.C.; ARAÙJO, E.L.; WiLLADiNo, L.G.; CAVALCANTE, U.M.T. (eds.). **Environmental stresses: Plant damage and benefits**. Recife: UFRPE: University Press, 2005.

LiMA, M.G. & GAsCoN, C. 1999. The conservation value of linear forest remnants in central Amazonia. **Biological Conservation** 91: 241-247.

LiMA, L.B. & MARCos FiLHo, J. 2009. Physiological conditioning of cucumber seeds and relationship with plant performance in the field. **Revista Brasileira de Sementes**, vol. 31, n° 3, p.027-037.

LiMA, L.B.; MARCos FiLHo, J. 2010. Physiological conditioning of cucumber seeds and germination under different temperatures. **Revista Brasileira de Sementes**, Brasilia, v. 32, n. 1, p. 138-147.

MARÇAL, J.A. 2011. **Initial growth of stone pine** (Jatropha curcas L.) **under irrigation with saline water in soil with organic matter**. 80 p. (Master's dissertation). Federal University of Paraiba.

MARCOS FiLHo, J. 2005. **Seed physiology of cultivated plants**. Piracicaba: FEALQ, p. 495.

MARCos FiLHo, J.; KiKUTi, J.A.D. 2008. Physiological conditioning of cauliflower seeds and plant performance in the field. **Horticultura Brasileira**, v.26, n.1, p.165169.

MAsCARENHAs, F.R.; MEDEiRos, D.C.D.; MEDEiRos, J.F.D.; DiAs,P.M.s., soUZA, M.S.D.M. 2010. Production and quality of Gâlia melon grown under different salinity levels. **Green Journal of Agroecology and Sustainable Development**, 5(5).

MEDEiRos J.F.; siLVA M.C.C.; sARMENTo, D.H.A.; BARRos A D. 2007. Growth of melon trees grown under different levels of salinity, with and without soil cover.

Revista Brasileira de Engenharia Agricola e Ambiental. v.11, n.3.

MEDEIROS, J.F.; DIAS, N.S.; BARROS, A.D. 2008. Irrigation management and melon tree tolerance to irrigation water salinity. **Revista Brasileira de Ciências Agràrias**. 3: 242247.

MEDEIROS, D.C.; MEDEIROS, J.F.; DIAS, P.M.S.; BARBOSA, M.A.G.; LIMA, K.S. 2010. Production of melon Pele de Sapo hybrid 'Medellin' grown with different water salinities. **Revista Ciência Agronòmica**. vol.42, n.3.

MEDEIROS, D.C.; MEDEIROS, J.F.; PEREIRA, F.A.L.; SOUZA, R.O.; SOUZA, P.A.

2011. Production and quality of cantaloupe melon grown with water of different salinity levels. **Revista Caatinga**. Vol. 24, n. 1.

MELO, R.M.; BARROS, M.F.C.; SANTOS, P.M.; ROLIM, M.M. 2008. Correction of saline-sodic soils by the application of mineral gypsum. **Brazilian Journal of Agricultural and Environmental Engineering.** 12.

MELO, A.D. **Operation of reservoirs in the semi-arid region considering water quality criteria**. 2005. 87 f. Dissertation (Master's Degree in Civil and Environmental Engineering) - Federal University of Campina Grande, Campina Grande, 2005.

MELONI, D.A. 2003. Photosynthesis and activity of superoxide dismutase, peroxidase and glutathione reductase in cotton under salt stress. **Environmental and Experimental Botany**, 49, 220-225.

MENDONÇA JÙNIOR, Antonio Francisco. 2015. **Growth, production and quality of melon and watermelon grown under *Ascophyllum nodosum* (L.) seaweed extract:** Antonio Francisco de

Mendonça Jùnior. - 126 f. Thesis (Doctorate) - Federal Rural University of the Semi-Arid. UFERSA. Mossoró.

MOING, A. AHARONI, A. BIAIS, B. 2010. Extensive metabolic cross-talk in melon fruit revealed by spatial and developmental combinatorial metabolomics. **New Phytologist**. 10: 1469-1475.

MORAIS, P.L.D.; SILVA, G.G.; MAIA, E.N.; MENEZES, J.B.. 2009. Evaluation of the post-harvest technologies used and the quality of noble melons produced for export. **Ciência e Tecnologia de Alimentos**, v.29, n.1, p.214-218.

MOREIRA, R.S; MELO, A.M.T; PURQUEIRO, L.F.V; TRANI, P.E; NARITA, N. 2010.

Melon (Cucumis melo). Available at: <http://www.infobibos.com/Artigos/2009_3/melao/index.htm>. Accessed on: November 23, 2014.

MOHAMMADI, G.R. 2009. The effect of seed priming on plant traits of late-spring seeded soybean (Glycine max L.). **Journal of Agriculture and Environmental Science**, 5: 322-326.

NASCIMENTO NETO, J.R.; Bomfim, G.V.; Azevedo, B.M.; Viana, T.V.A.; Vasconcelos, D.V. 2012. Forms of application and doses of nitrogen for the yellow melon tree on the coast of Cearà. **Irriga**, v.17, p.364-375.

NASCIMENTO, W.M; SILVA, J.B.C.; SANTOS, E.C.; RICARDO, C. 2009. Germination of carrot seeds osmotically conditioned and pelletized with different ingredients. **Horticultura Brasileira**, Brasilia, v. 27, n. 1, p. 12-16.

NASCIMENTO, W.M.; WEST, S.H. 2000. Drying during muskmelon (*Cucumis melo* L.) seed priming and its effects on seed germination and deterioration. **Seed Science and Technology**, 28: 211-215.

NUNES, G.H.S.; SANTOS JÛNIOR, J.J.S.; ANDRADE, F.V.; BEZERRA NETO, F.; ALMEIDA, A.H.B.; MEDEIROS, D.C. 2004. Productive and quality aspects of melon hybrids grown in the Mossoró-Assu agropolo. **Horticultura Brasileira**, Brasilia, v. 22, n. 4.

OLIVEIRA, F.A.; MEDEIROS, J.F.; OLIVEIRA, M.K.T.; LIMA, C.J.G.S.; GALVÂO, D.C. 2008. Effect of saline water on the germination of Stylosanthes capitata Vogel. **Revista Verde de Agroecologia e Desenvolvimento Sustentàvel**, v.3, p.77-82.

PACHECO, M.V.; FERRARI, C.E.S.; BRUNO, R.L.A.; ARAÙJO, F.S.; SILVA, G.Z.; ARRUDA, A. A. 2012. Germination and vigor of Capparis flexuosa L. seeds subjected to salt stress. **Revista**

Brasileira de Ciências Agrârias, v. 7, n.2.

PANDITA, V.K.; ANAND, A.; NAGARAJAN, S.; SETH, R.; SINHA, S.N.. 2010. Solid matrix priming improves seed emergence and crop performance in okra. **Seed Science and Technology**, Zurich, v. 38, n. 3, p. 665-674.

PEREIRA, A.M.; QUEIROGA,R.C.F.; SILVA, G.D.; NASCIMENTO, M.G.R.; ANDRADE, S.E.O. 2012. Germination and initial growth of melon submitted to seed osmoconditioning with NACl and water salinity levels. **Revista Verde,** Mossoró - RN, v. 7, n. 3, p. 205-211.

PEREIRA, M.D.; DIAS, D.C.F.S.; DIAS, L.A.S.; ARAÙJO, E.F. 2008. Germination and vigor of carrot seeds osmoconditioned in moistened paper and aerated solution. **Revista Brasileira de Sementes**, Londrina, v.30 n.2, p. 137-145.

PIRES, R.C.M.; SAKAI, E.; ARRUDA, F.B.; FOLEGATTI, M.V. 2001. Crop water requirements and irrigation management. Piracicaba: **FUNEP**. Agricultural Engineering Series, 1.

PURQUERIO, L.F.V.; CECìLIO FILHO, A.B. Nitrogen concentration in the nutrient solution and number of fruits on melon fruit quality. **Horticultura Brasileira**, Brasilia, v.23, n.3, p.831-836, 2005.

QUEIROGA, R.C.F.; ANDRADE NETO, R.C.; NUNES, G.H.S.; MEDEIROS, J.F.; ARAÙJO, W.B.M. 2006. Germination and initial growth of melon hybrids as a function of salinity. **Horticultura Brasileira**, v. 24, p. 315-319.

RAI, A.K.; BASU, A.K. 2014. Pre-sowing seed bio-priming in okra: response for seed production. **The Bioscan, Ranchi**, v. 9, n. 2, p. 643-647.

REIS, C.F.; VILAS BOAS, M.A.; MERCANTE, E.; HERMES, E.; REISDORFER, M. 2011.

Evaluation of water quality for irrigation in Salto do Lontra-PR. **Engenharia Ambiental, Espirito Santo do Pinhal**, v. 8, n. 1.

REIS, R.G.E.; GUIMARÂES, R.M.; PEREIRA, D.S.; CASTRO, M.B.; VIEIRA, A.R.; CARVALHO, M.L.M. 2013. Physiological quality of osmoconditioned eggplant seeds subjected to drying. **Pesquisa Agropecuâria Brasileira**, Brasilia, v. 48, n. 11, p. 1507-1516.

RIBEIRO, G.M.; MAIA, C.E.; MEDEIROS, J.F. 2005. Use of linear regression to estimate the relationship between electrical conductivity and the ionic composition of irrigation water. **Revista Brasileira de Engenharia Agricola e Ambiental**, 9: 13-22.

RODRIGUES, A.P.D.C.; LAURA, V.A.; CHERMOUTH, K.S.; GADUN,J. 2009. Osmoconditioning of parsley seeds (Petroselinum sativum Hoffm.) at different water potentials. **Ciências Agrârias, Lavras**, v. 33 n.5, p.1288-1294.

SENHOR, R.F; ANDRADE NETO, R.C.; SOUZA, P.A.; MENEZES, J.B.; MATOS, D.S.S. 2008. Refrigerated storage of yellow melon hybrid Frevo grown during the rainy season. **Caatinga**, Mossoró, v. 21, n. 2.

NATIONAL RURAL APPRENTICESHIP SERVICE. 2007. **Growing melons: handling, harvesting, post-harvesting and marketing**. Brasilia: SENAR, 104 p. (SENAR-131 collection).

SILVA, P.S.; MENEZES, J.B.; OLIVEIRA, O.F.; SILVA, P.I.B. 2004. Distribution of Total Soluble Solids in Melon. **Horticultura Brasileira, Brasilia**, v. 21, n. 1, p. 31-33, mar.

SILVA, M.C.C.; MEDEIROS, J.F.; NEGREIROS, M.Z.; SOUSA, V.F. 2005. Melon fruit yield under different levels of irrigation water salinity, with and without soil cover. **Horticultura Brasileira, Brasilia**, v.23, n.2, p.202-205.

SILVA, M.V.T.; SOUSA LIMA, R.M.; OLIVEIRA, F.L.; SILVA, N.K.C.; MEDEIROS, J.F. 2014. Zucchini production under different levels of saline water and nitrogen doses. **Green Journal of Agroecology and Sustainable Development**, 9(1), 278-284.

SIQUEIRA W.C. et al. 2009. Fruit quality of yellow melon grown in a greenhouse under different irrigation rates. **Agrotechnological Science,** 33: 1041-1046.

SINGH, P.K. et al. 2014. Germination and field emergence in osmotic and solid matrix priming in onion (*Allium cepa*). **Indian Journal of Agricultural Sciences**. New Delhi, v. 84, n. 12. '; 1561-1564.

SIVITREPE, N; SIVITREPE, HO; ERIS, A. 2003. The effect of NaCl priming on salt tolerance in melon seedling grown under saline conditions. **Scientiae Horticulturae,** 97: 229237.

TAIZ, L. ZEIGER, E. 2013. **Plant Physiology**. 5. ed. Artmed, Porto Alegre.

TAYER, J.N. 1987. **Effect of irrigation methods with saline water on maize production and salt distribution in soil.** 285 p. (Doctor of thesis). Bari: International Center for Advanced Mediterrane an Agronomic Studies).

TERCEIRO NETO C. P.C.T.; MEDEIROS, J.F.; GHEYIH, R.; DIAS N.S.; OLIVEIRA, F.R. 2014. Growth and mineral composition of 'Pele de sapo' melon plant tissue under saline water management.

Irriga, Botucatu, v. 19, n. 2.

TERCEIRO NETO, C.P.C.; GHEYI, H.R.; MEDEIROS, J.F.; DIAS, N.S.; CAMPOS, M.S. 2013. Melon productivity and quality under management with water of increasing salinity.

Pesquisa Agropecuària Tropical, Goiânia, v. 43, n. 4, p. 354-362.

TESTER, M.; DAVENPORTE, R. 2003. Na tolerance and Na+ transport in higher plants. **Annals of Botany**, 91: 503-527.

TOMAZ, H.V.Q.; AROUCHA, E.M.M.; NUNES, G.H.S.; BEZERRA NETO, F.; TOMAZ, H.V.Q.; QUEIROZ, R.F. 2009. Post-harvest quality of different melon hybrids - Amarelo stored under refrigeration. **Revista Brasileira de Fruticultura. Jaboticabal**, v. 31, n. 4.

VALE, M.F.S. **Pruning and planting density in melon hybrids**. 2000. 41f. Dissertation (Master's Degree in Plant Science) - Mossoró School of Agriculture, Mossoró.

VENANCIO, D.F.V.; SANTOS, R.M.; CASSARO, S.; PIERRO, P.C.C. 2015. The water crisis and its global context. **Enciclopédia biosfera**, Centro Cientifico Conhecer - Goiânia, v.11 n.22; p.2.

VON SPERLING, M. 2005. **Introduction to water quality and sewage treatment. Belo Horizonte, UFMG, p. 425.**

WARTERWORTH, W.M. DRURY, G.E.; BRAY, C.M.; WEST, C.E. 2011. Repairing breaks in the plant genome: the importance of keeping it together. **New Phytologist**, Cambridge, v. 192, n. 4, p. 805-822.

ZONG, L. TEDESCHI, A.; XUE, X.; WANG, T.; MENENTI, M.; HUANG, C. 2011. Effect of different irrigation water salinities on some yield and quality components of two field- grown cucurbits species. **Turkish Journal of Agriculture and Forestry**, Ankara, 35: 25-35.

APPENDICES

Appendix A - Summary of analysis of variance for mean values of fruit mass, number of fruits per plant and total yield of melon as a function of seed treatment with NaCl. CCTA/UFCG, Pombal - PB, 2016.

Sources of Variation	GL	Mean Squares		
		MFRUT	NFP	EN
BLOCK	3	0.11^{01ns}	$0{,}14^{-01ns}$	$18{,}84^{ns}$
TRATSE	1	0.10*	0,15*	198,61*
ERROR (A)	3	0.58^{-02ns}	$0{,}26^{-01ns}$	$17{,}59^{ns}$
NIVELSA	3	0.70^{-02ns}	0,22*	67,07*
NIVELSA*TRATSE	3	0.77^{-03ns}	0,25*	42,17*
ERROR (B)	18	0.85^{-02}	0,24	8,65
C.V. (%)	9.68	9,68	6,53	10,33

(*) Significant at 5% probability and (ns) not significant by the F Test.

Appendix B - Summary of analysis of variance for mean values of length, diameter, shape index, firmness and pulp thickness of melon fruit as a function of seed treatment with NaCl. CCTA/UFCG, Pombal - PB, 2016.

Sources of Variation	GL	Mean Squares				
		COMP	DIAM	IFF	FIRMPOL	ESPMESO
BLOCK	3	2,19*	1,49*	$0{,}16^{ns}$	12,05*	$0{,}89^{ns}$
TRATSE	1	$0{,}16^{ns}$	$0{,}84^{ns}$	$0{,}23^{ns}$	$6{,}20^{ns}$	$0{,}20^{ns}$
ERROR (A)	3	$0{,}24^{ns}$	$0{,}40^{ns}$	$0{,}60^{ns}$	$2{,}89^{ns}$	$0{,}30^{ns}$
NIVELSA	3	1,38*	$0{,}48^{ns}$	0,67*	$6{,}73^{ns}$	0,71*
NIVELSA*TRATSE	3	$0{,}46^{ns}$	$0{,}15^{ns}$	$0{,}43^{ns}$	$0{,}99^{ns}$	$0{,}11^{ns}$
ERROR (B)	18	0,39	0,24	0,57	5,16	0,20
C.V. (%)	9.68	4,15	3,54	2,18	9,05	3,43

(*) Significant at 5% probability and (ns) not significant by the F Test.

Appendix C - Summary of analysis of variance for mean values of soluble solids, total acidity and ripeness index of melon fruit as a function of seed treatment with NaCl. CCTA/UFCG, Pombal - PB, 2016.

Sources of Variation	GL	Mean Squares		
		SST	ATT	SS/AT
BLOCK	3	0,60^{ns}	0,51^{ns}	6142,349^{ns}
TRATSE	1	0,48^{ns}	0,20^{ns}	2865,204^{ns}
ERROR (A)	3	1,61^{ns}	0,35^{ns}	2934,018^{ns}
NIVELSA	3	0,67^{ns}	0,84^{ns}	1120,351^{ns}
NIVELSA*TRATSE	3	0,16^{ns}	0,28^{ns}	36,250^{ns}
ERROR (B)	18	0,98	0,51	440,923
C.V. (%)	9.68	11,29	23,18	20,69

(*) Significant at 5% probability and (ns) not significant by the F Test.

More Books!
yes

MIX
Papier aus verantwortungsvollen Quellen
Paper from responsible sources
FSC® C105338

Printed by Books on Demand GmbH, Norderstedt / Germany